WHAT DO YOU SEE?

By

Michael C. Carlson

1stBooks – rev. 01/20/04

Ancient people from around the world have gazed into these stones and saw spirits. What do you see?

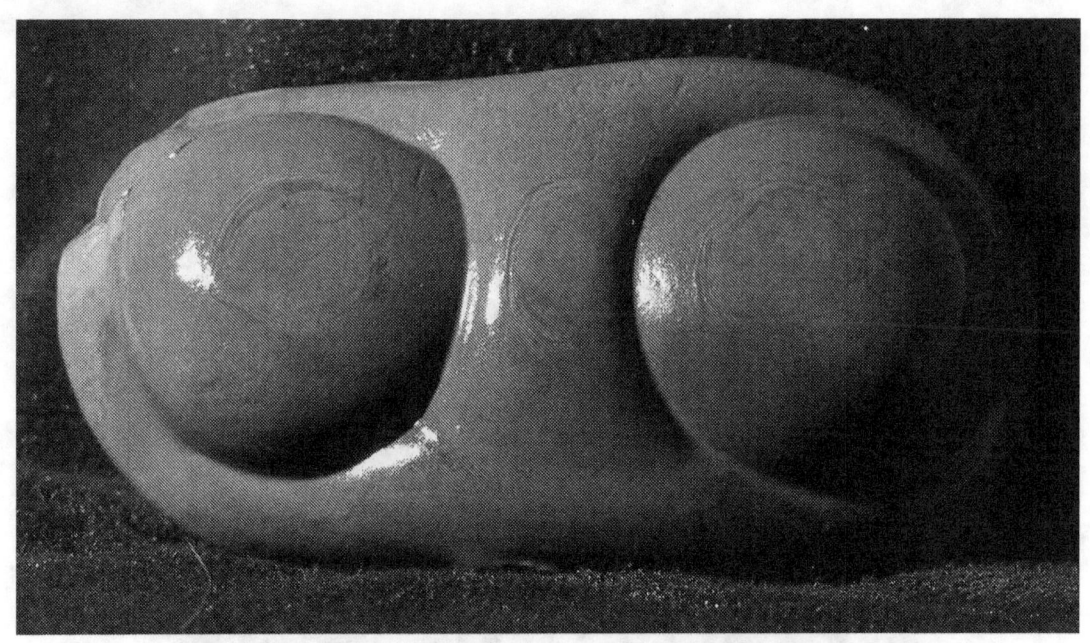

Michael C. Carlson

"Spirit Stones" grow in the ground where mineral-laden water seeps over the top of a layer of clay, a true rock garden.

Known as "concretions" by geologists, they form around a nucleus of matter that attracts the minerals to a central point giving the stones much of their rounded shapes.

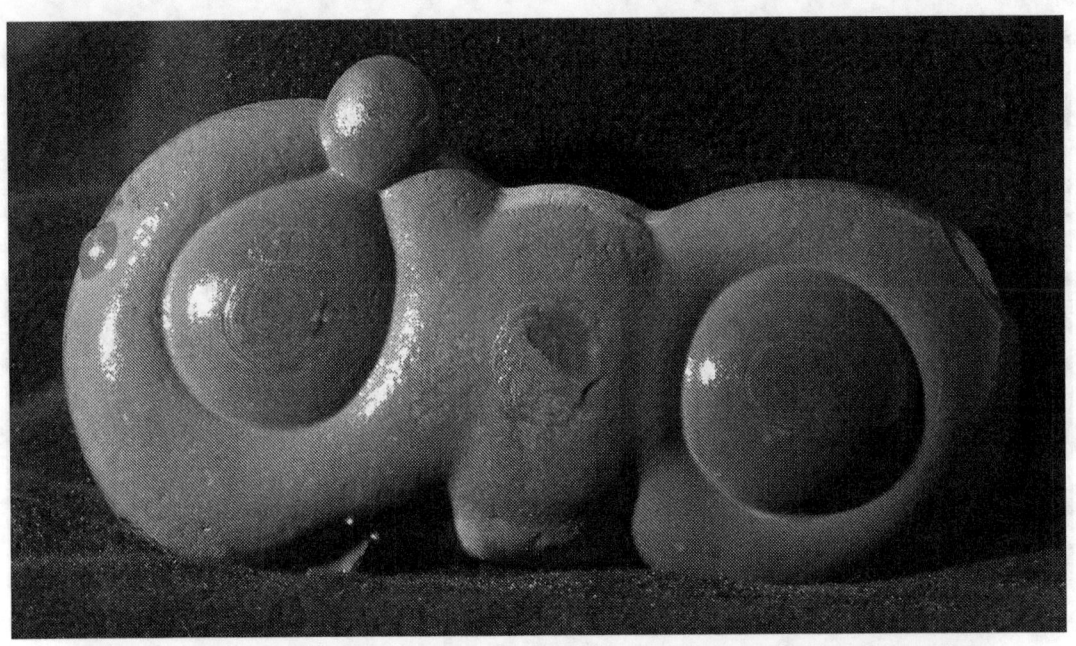

Michael C. Carlson

Growing like stalactites in the earth until something disturbs them, they are usually found in rivers, along beaches, and where glaciers have retreated.

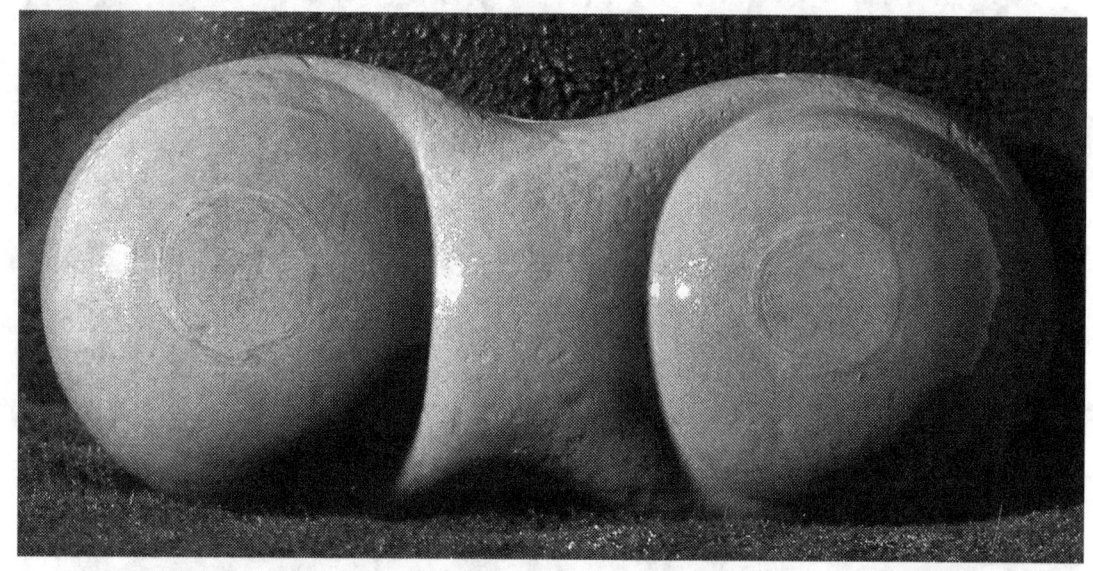

There is much evidence ancient people truly believed there were spirits inside of them. Many called them "Spirit Stones." This is your chance to see what they were looking at.

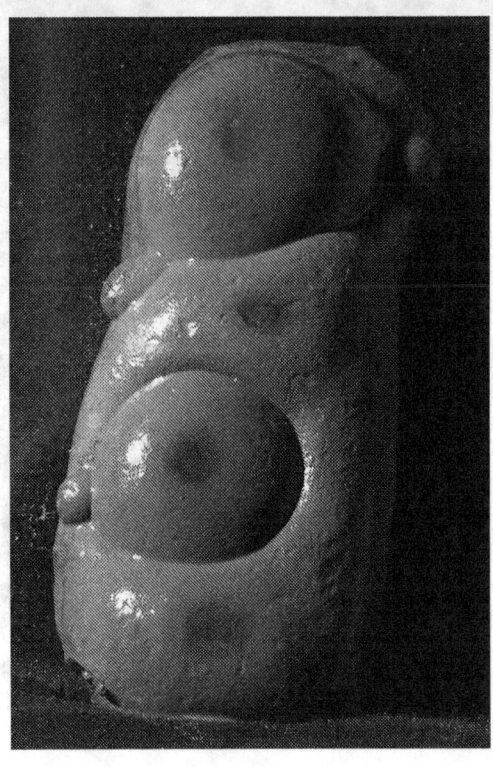

Michael C. Carlson

These stones are found on the beaches of the Kenai Peninsula, Alaska.

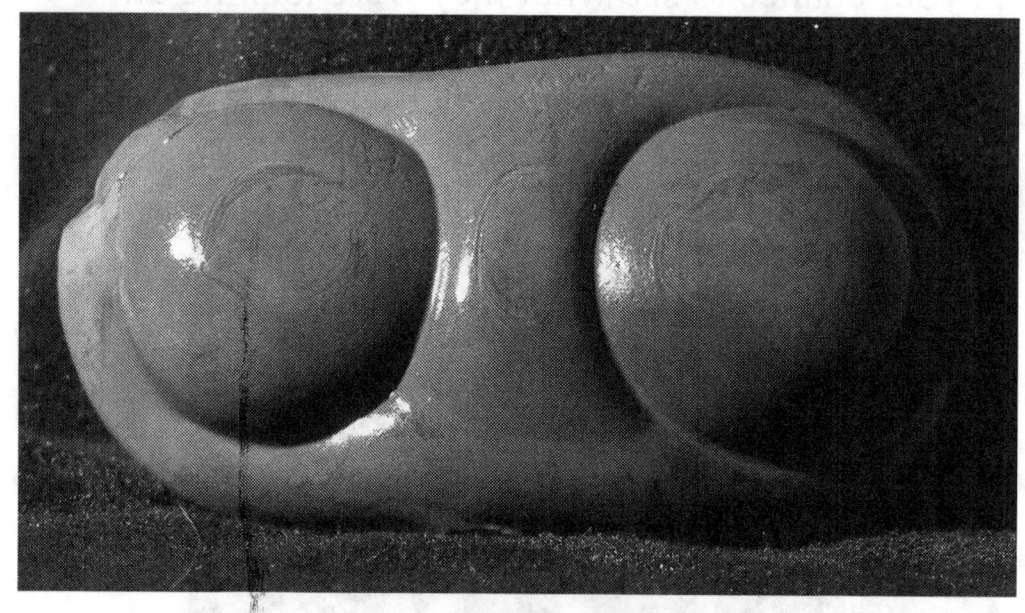

The backsides look like a turtle's shell and they "grow" face down.
Contributors of stones: Liza Wondra, Charlie, Reuben, Curtis, Susan, and Mike Carlson.

Michael C. Carlson

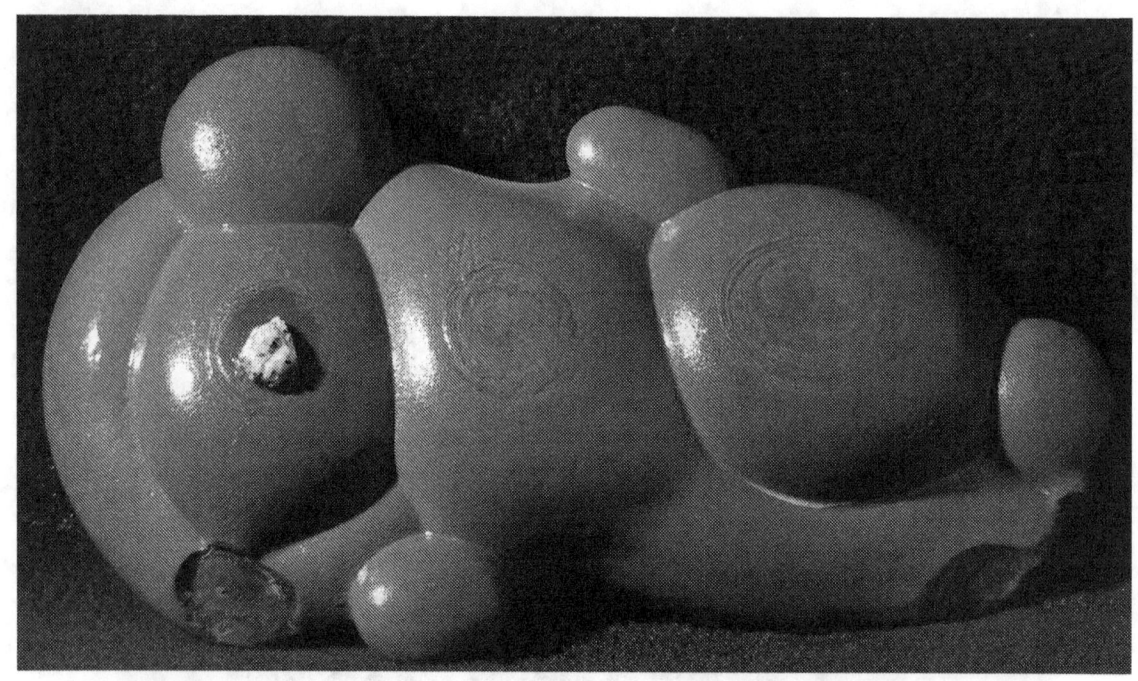

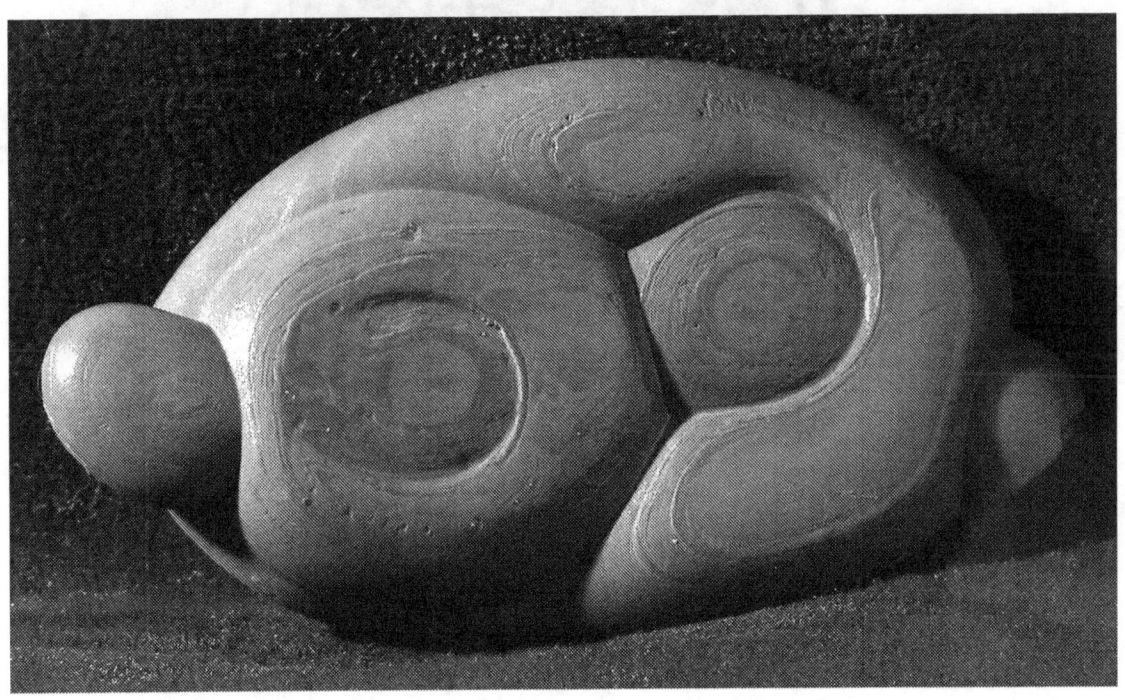

Michael C. Carlson

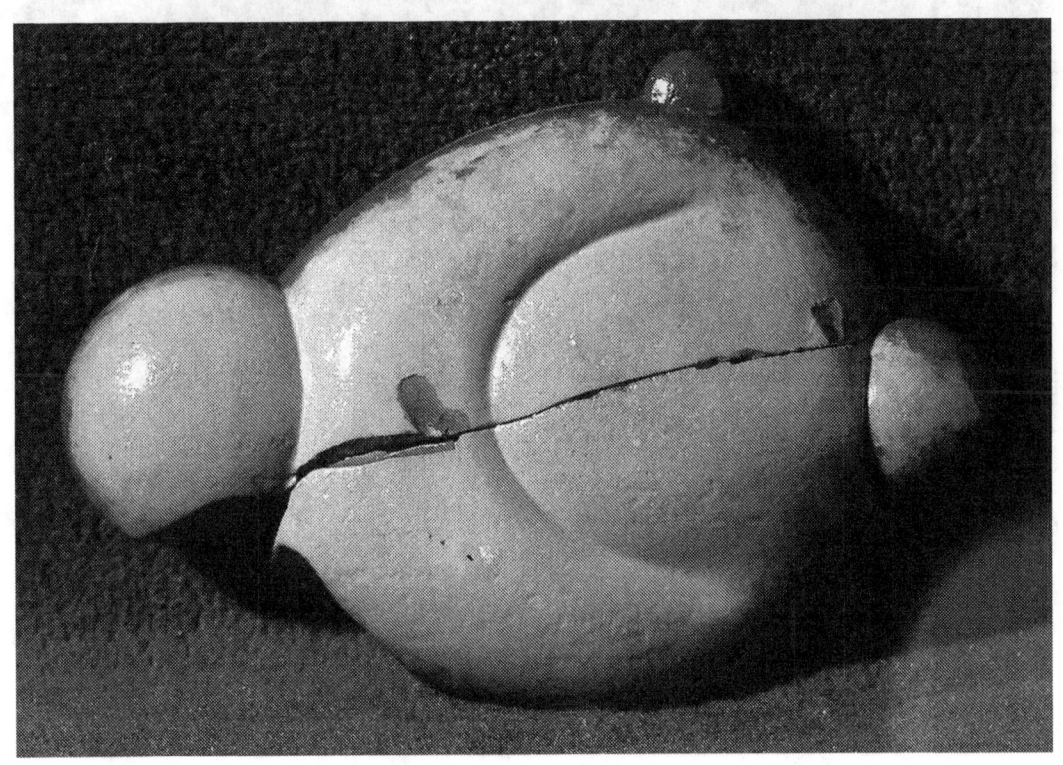

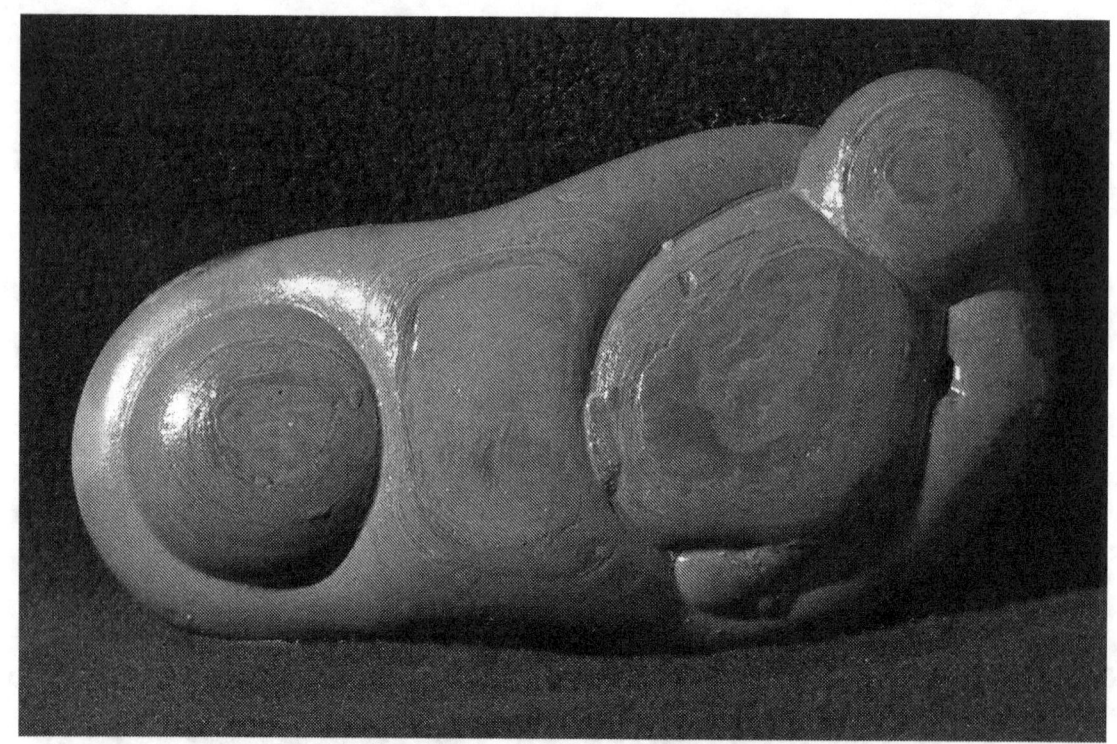

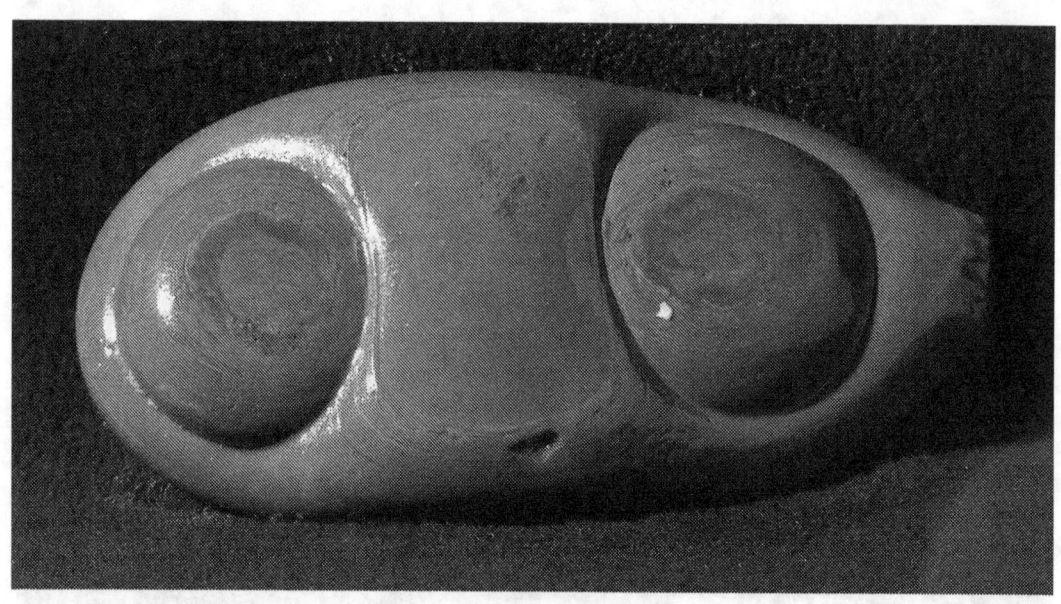

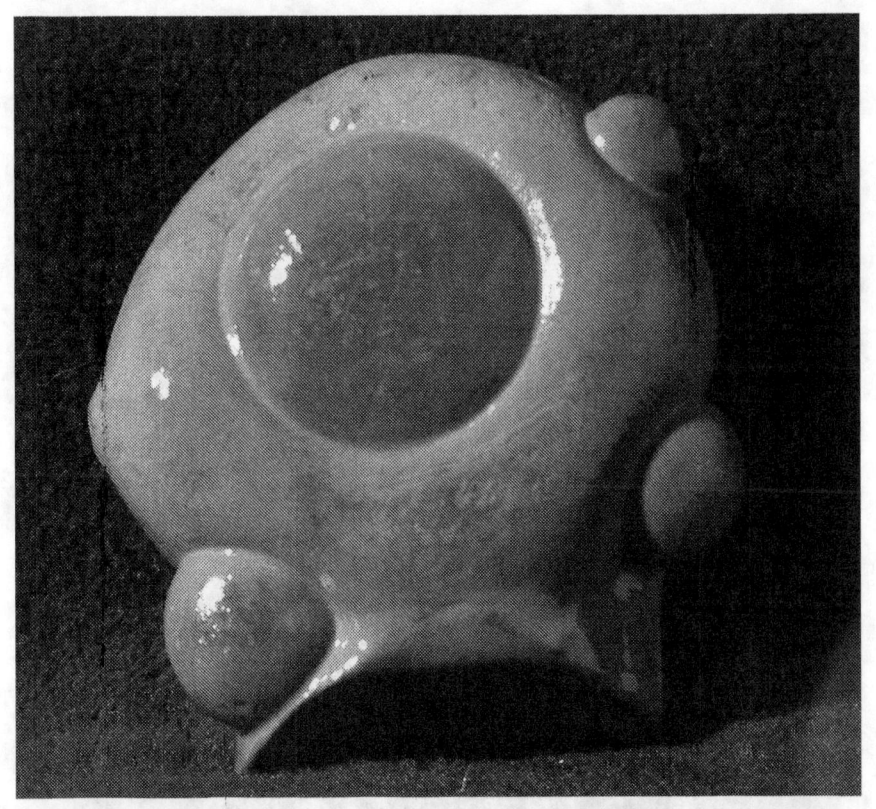

Michael C. Carlson

Michael C. Carlson

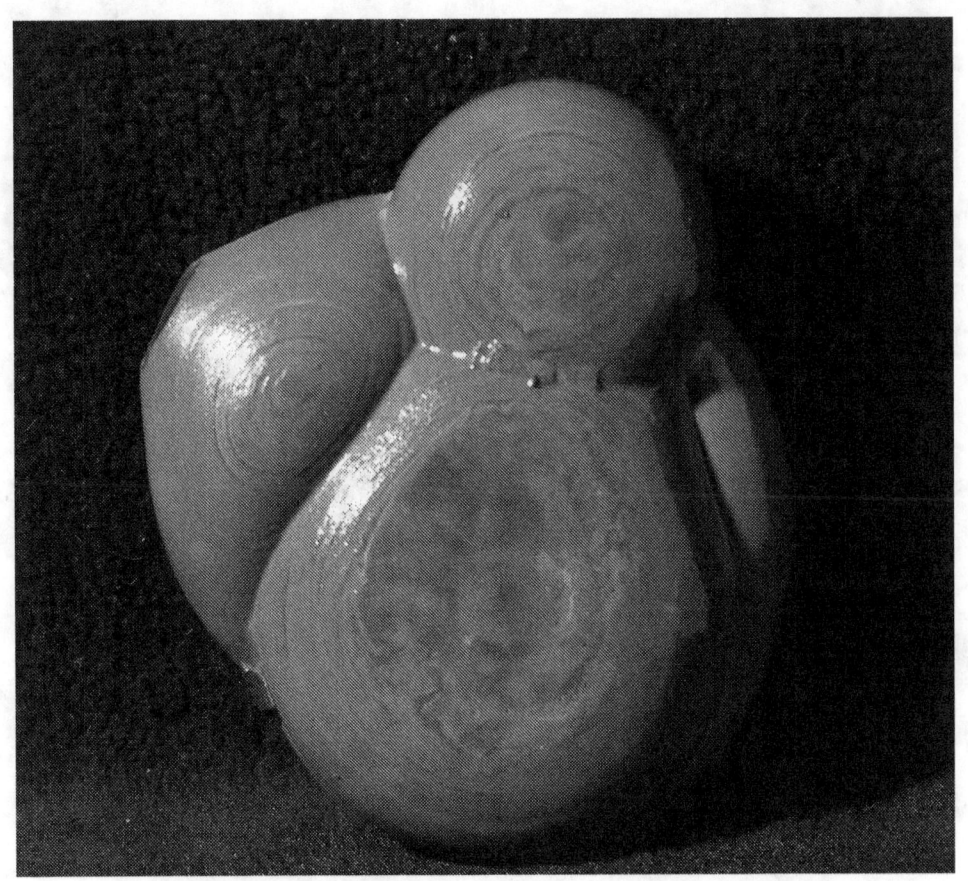

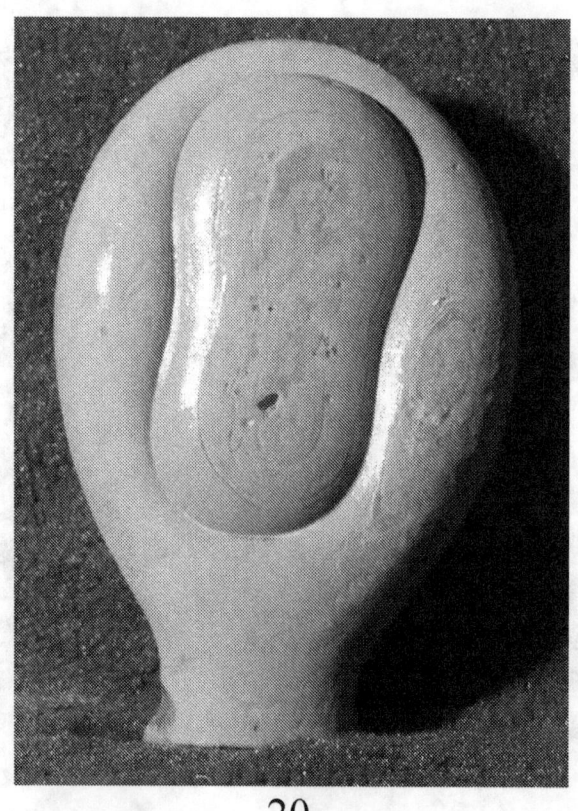

Michael C. Carlson

23

Michael C. Carlson

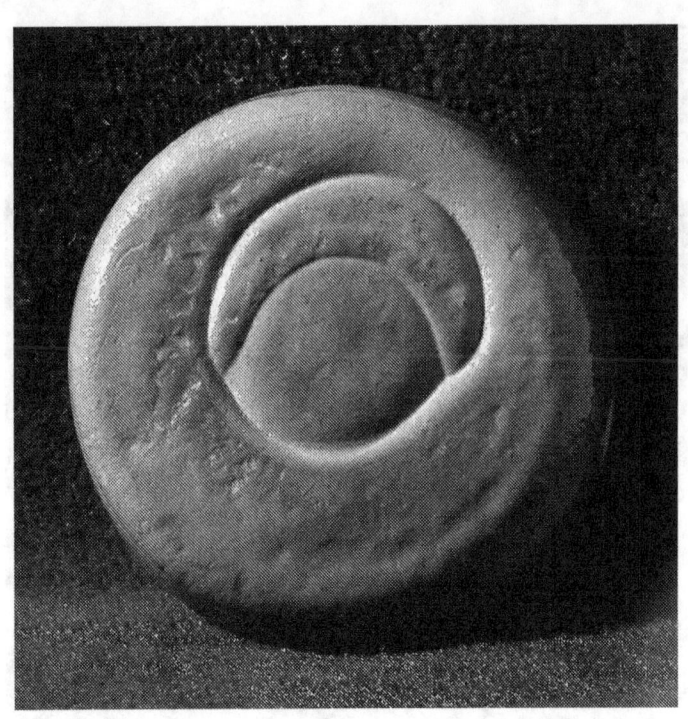

Michael C. Carlson

27

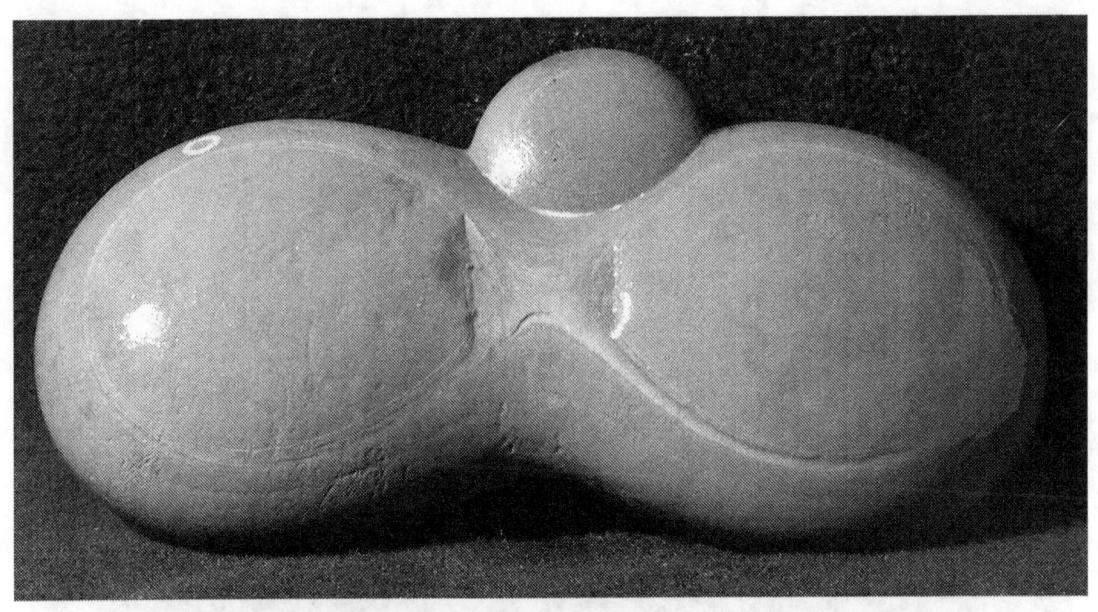

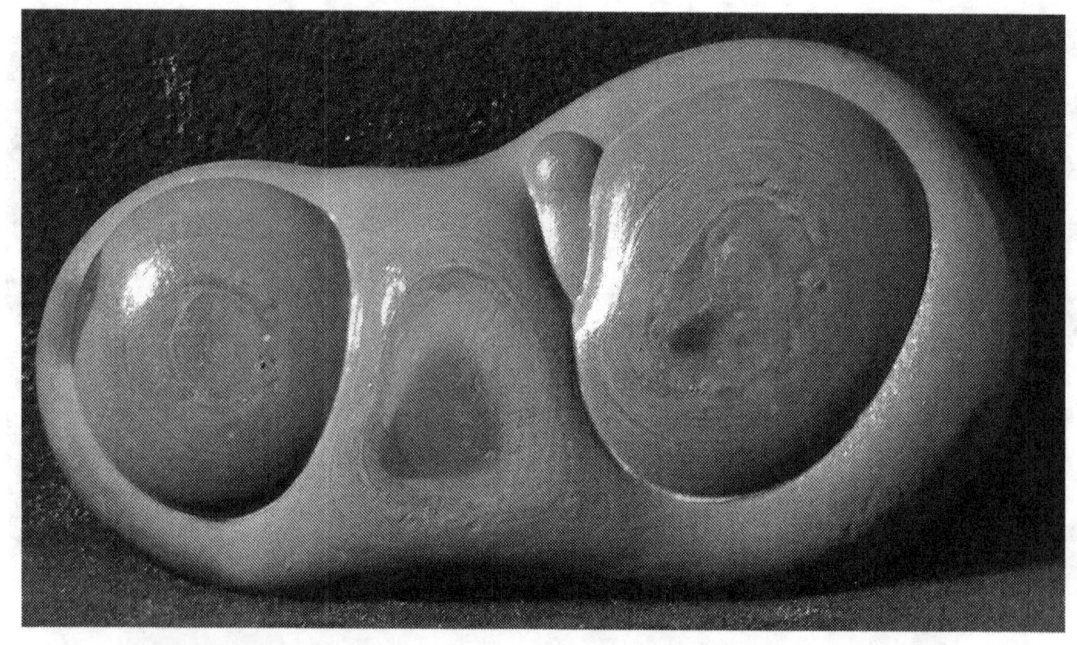

Michael C. Carlson

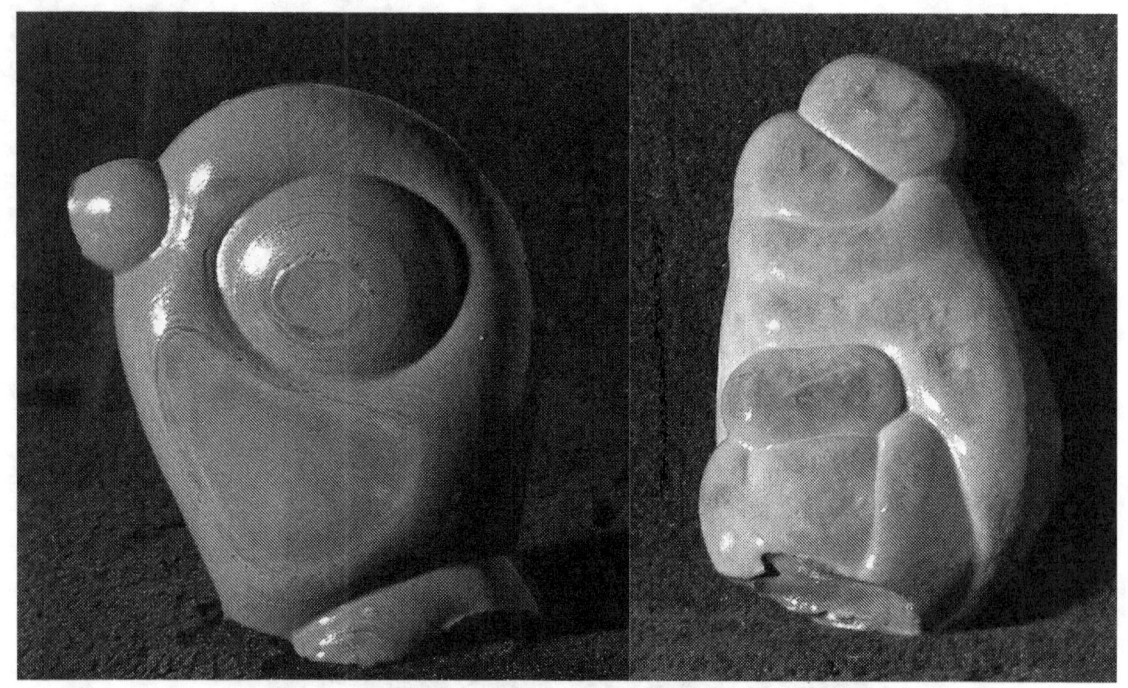

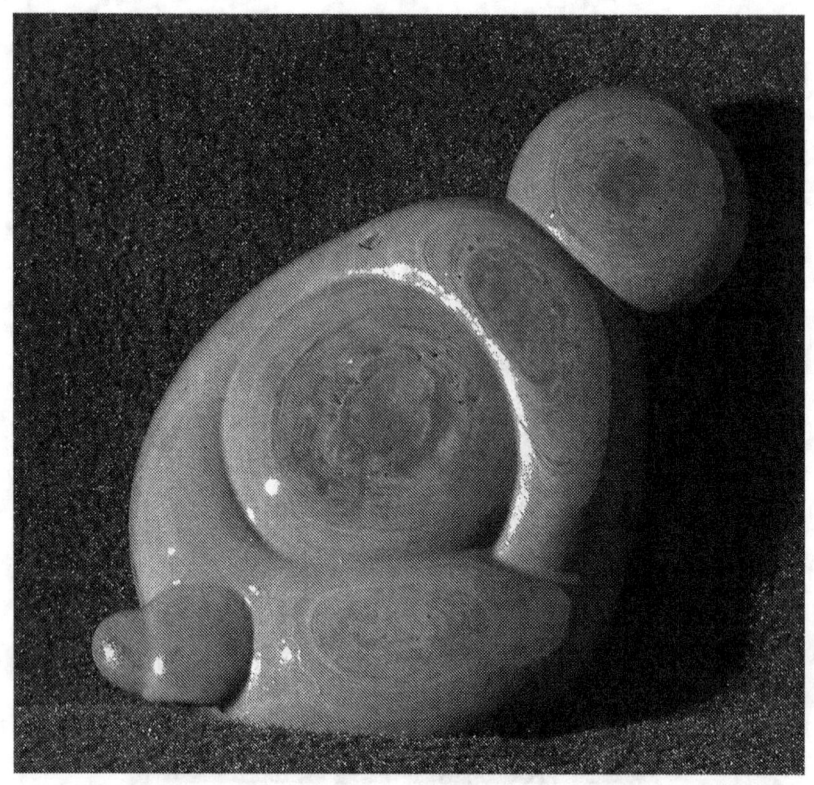

Michael C. Carlson

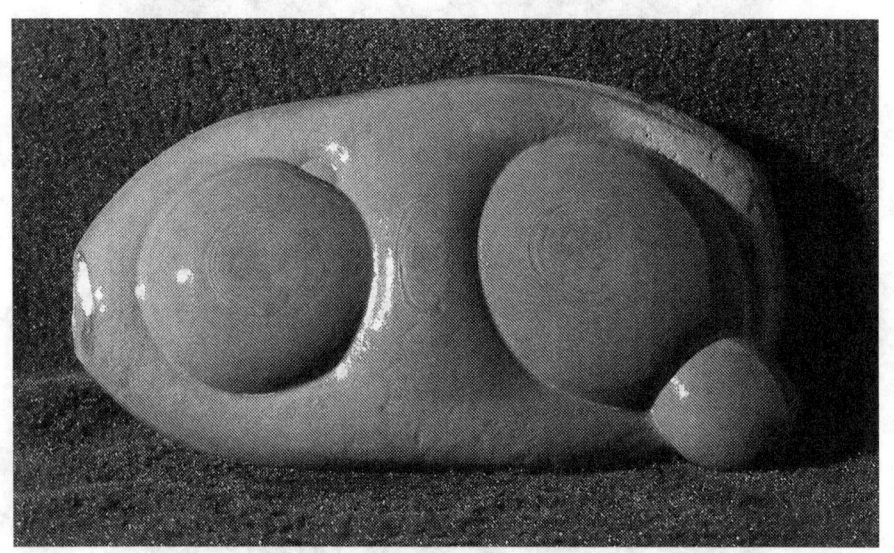

49

Michael C. Carlson

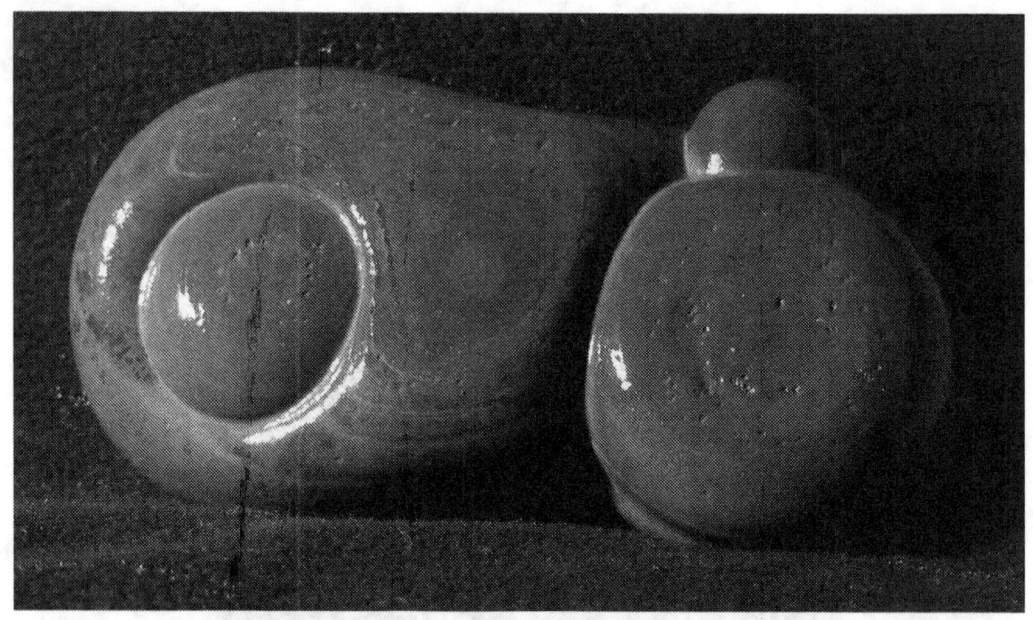

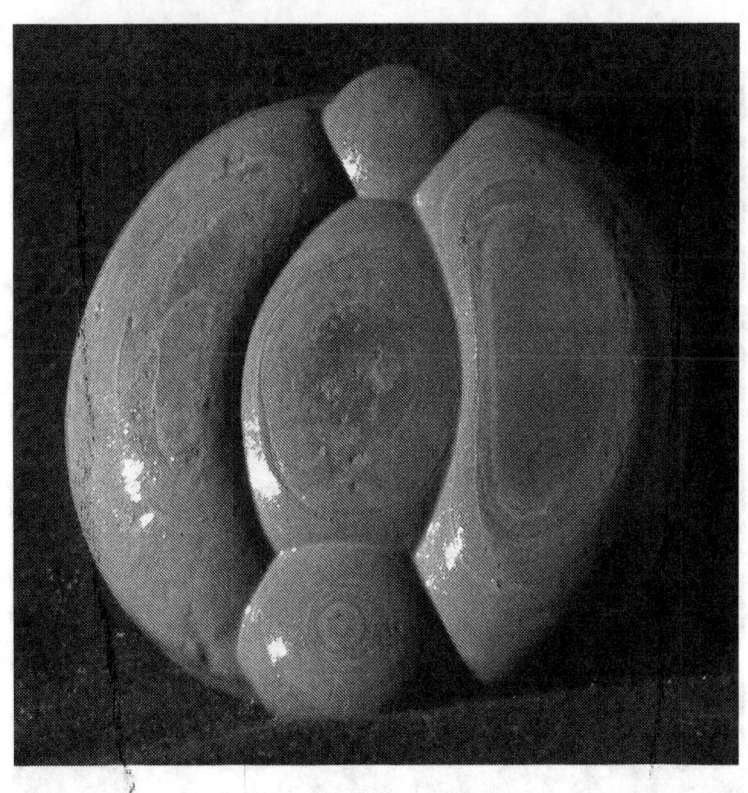

51

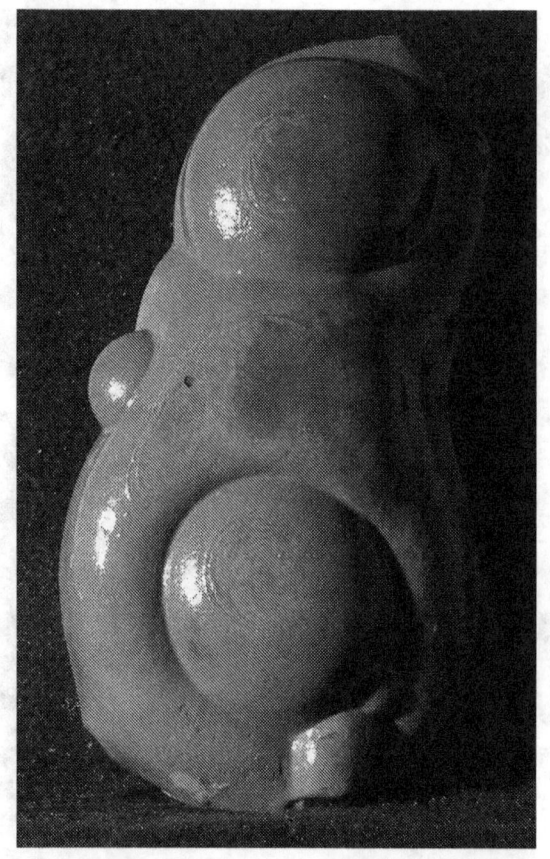

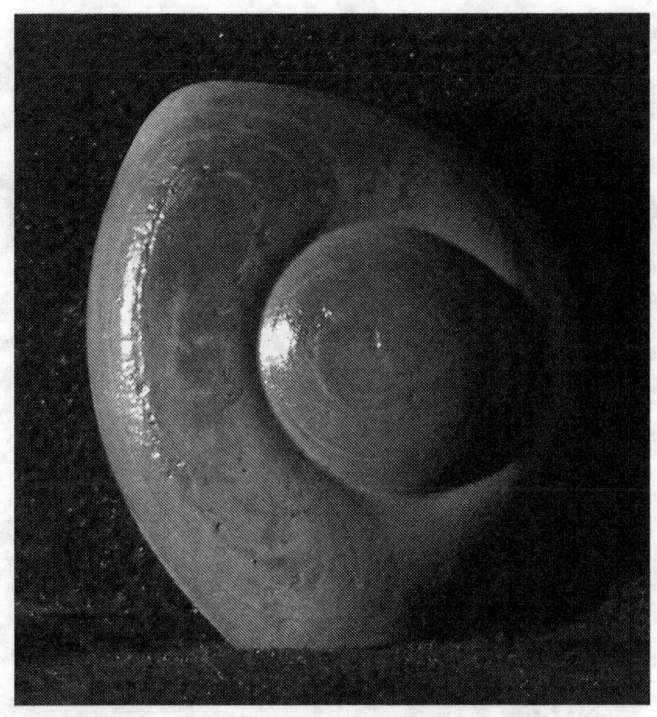

53

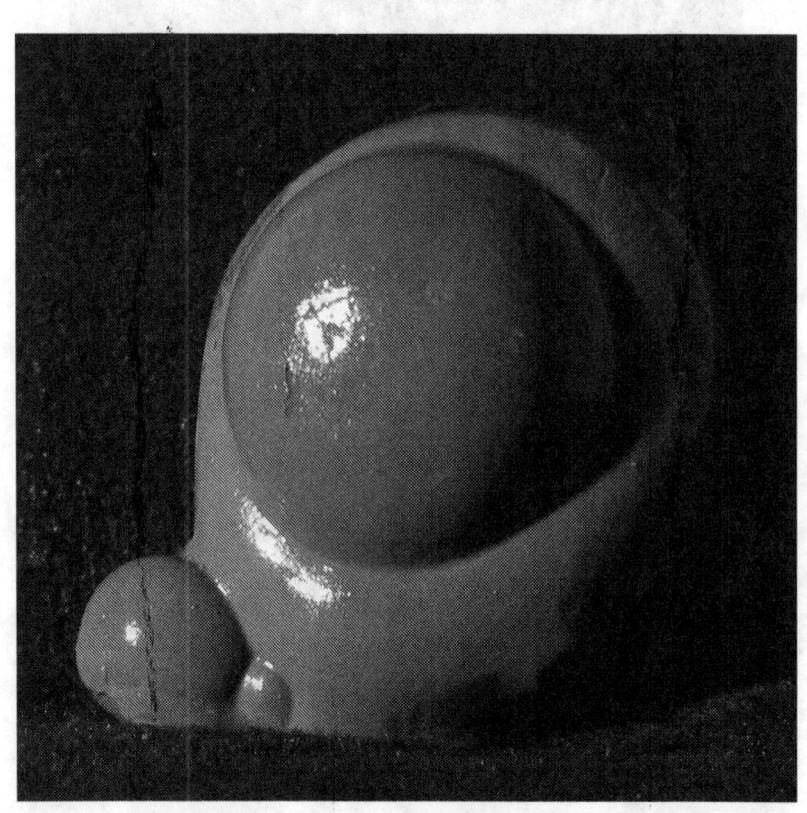

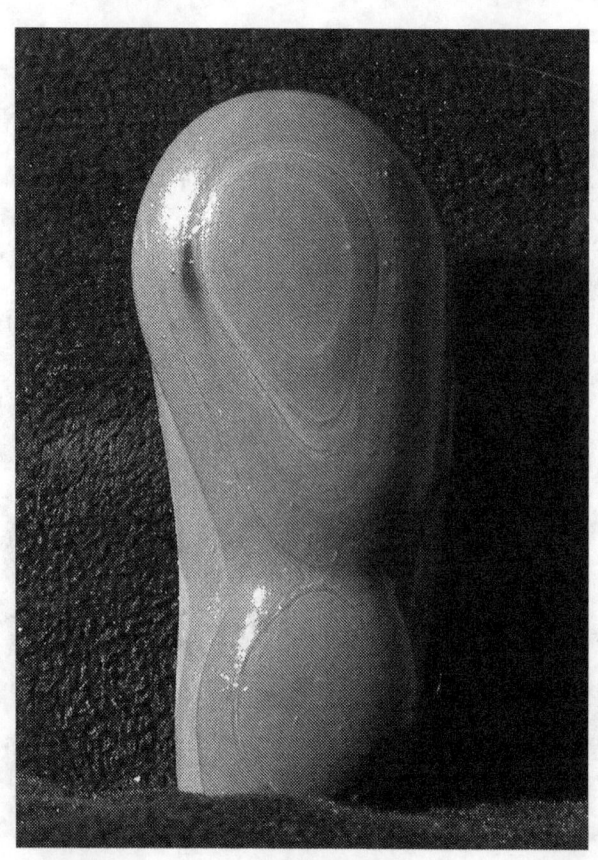

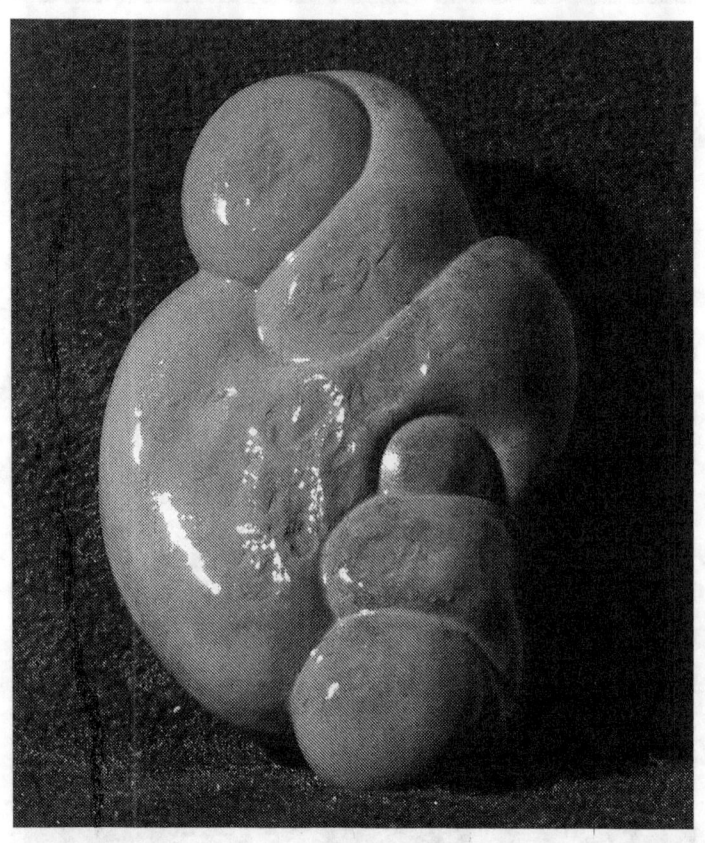

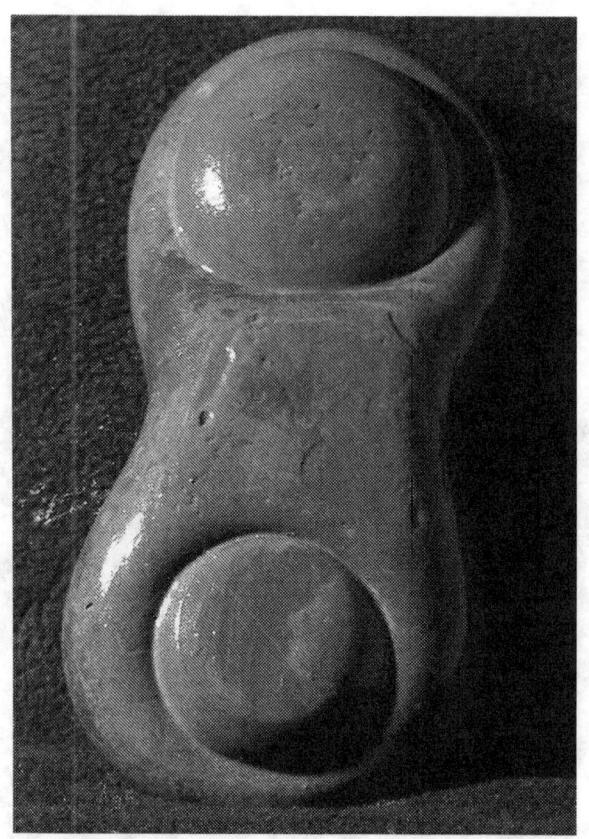

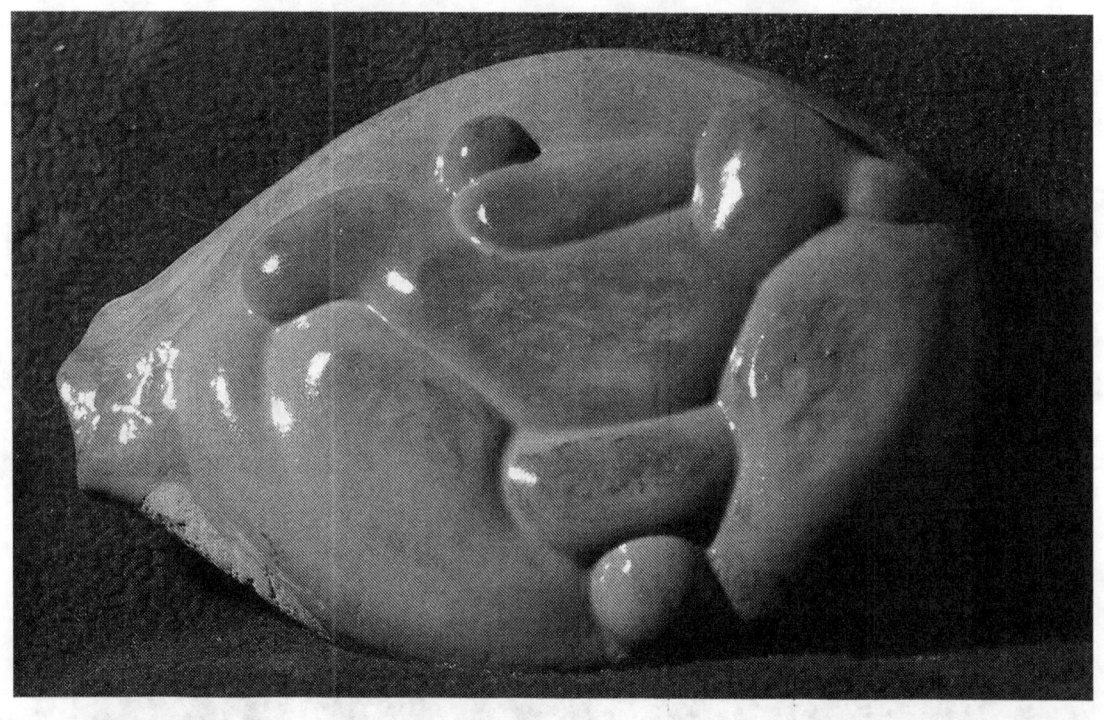

Michael C. Carlson

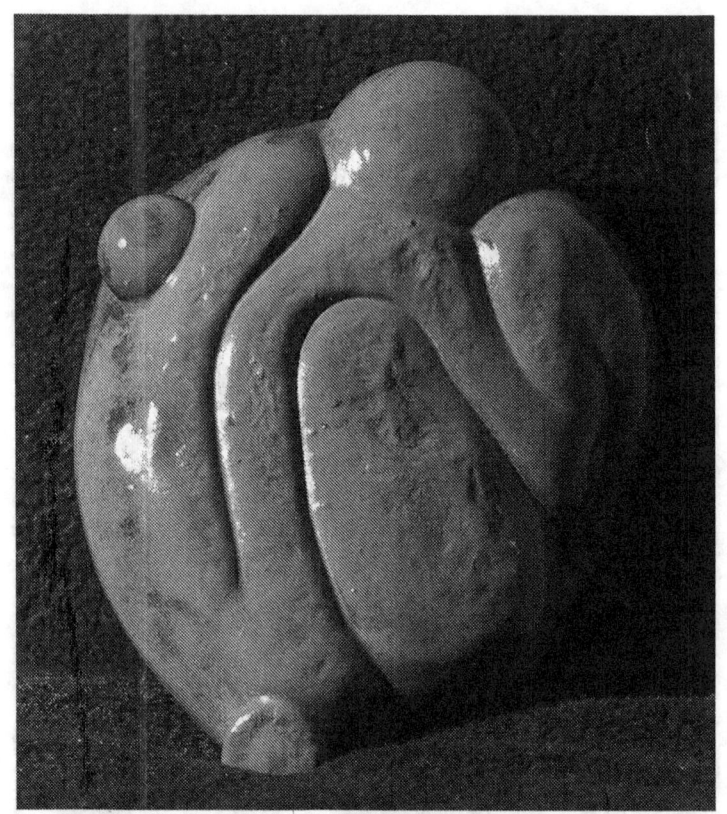

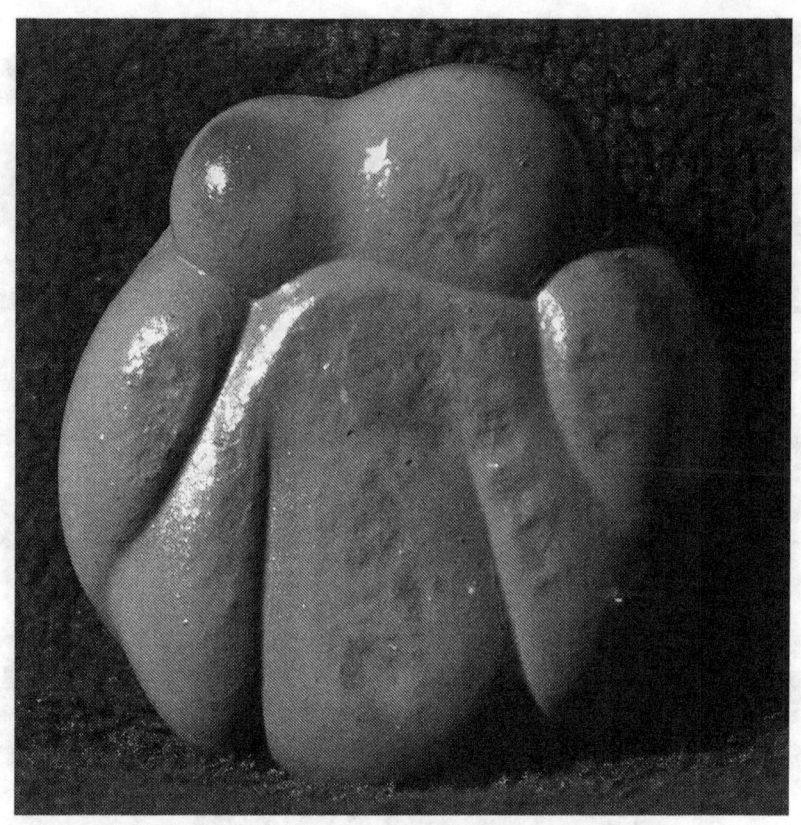

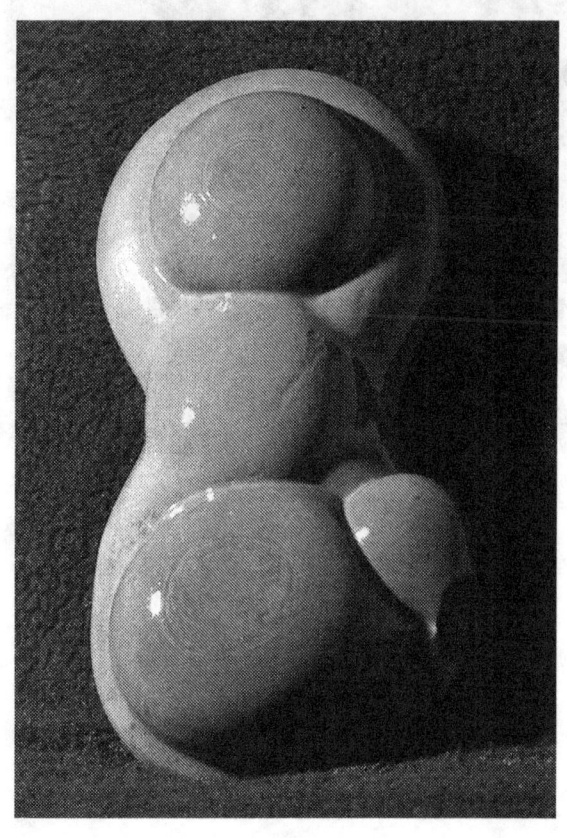

Michael C. Carlson

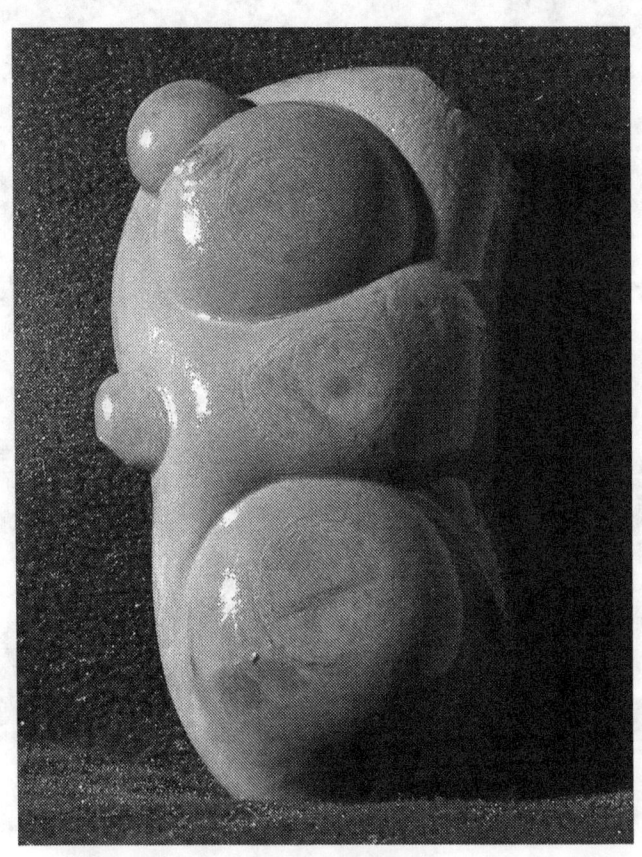

About the Author

Mike Carlson has been beachcombing Alaskan beaches on and off for twenty years. These stones are the "Darndest things he ever saw". This book is his effort to share them with you.